Elapsed Time at the Olympics

Dianne Irving

Real World Math Books are published by Capstone Press,
151 Good Counsel Drive, P.O. Box 669, Mankato, Minnesota 56002.
www.capstonepub.com

Printed in the United States of America in North Mankato, Minnesota.
032010
005740CGF10

Books published by Capstone Press are manufactured with paper containing at least 10 percent post-consumer waste.

Library of Congress Cataloging-in-Publication Data
Irving, Dianne.
Elapsed time at the Olympics / by Dianne Irving.—1st hardcover ed.
p. cm.—(Real world math level 4)
Includes index.
ISBN 978-1-4296-5239-1 (library binding)
1. Sports—Technological innovations—Juvenile literature.
2. Olympics—Records—Juvenile literature. 3. Time measurements—Juvenile literature. 4. Problem solving—Juvenile literature. I. Title. II. Series.
GV745.L63 2011
688.7′6—dc22 2010001818

Editorial Credits
Sara Johnson, editor; Emily R. Smith, M.A.Ed., editorial director; Sharon Coan, M.S.Ed., editor-in-chief; Lee Aucoin, creative director; Rachelle Cracchiolo, M.S.Ed., publisher

Photo Credits
The author and publisher would like to gratefully credit or acknowledge the following for permission to reproduce copyright material: cover Photolibrary.com; p.1 AAP Image; p.4 Photolibrary.com/Bridgeman Library; p.5 Getty Images; p.6 (both) Photolibrary.com; p.7 Corbis/Richard Cohen; p.8 (left) Corbis; p.8 (right) Northwind; p.9 (top) Shutterstock; p.9 (below) Corbis/Gideon Mendel; pp.10-11, Photolibrary.com; p.11 (inset) AAP Image/Russell McPhedran; p.12 AAP Image/Kim Ludbrook; p.13 Getty Images; p. Photolibrary.com; p.15 Corbis/Tom Stewart; p.16 (all) Shutterstock; p.17 Getty Images/Phil Walter; p.18 AAP Image/Rick Rycroft; p.19 Getty Images/Yoshikazu Tsuno; p.20 Corbis/Adrian Bradshaw; p.21 Corbis/Carl Purcell; p.22 (second from bottom) Photolibrary.com; p.22 (all others) Shutterstock; p.23 Getty Images/Matt Turner; p.24 Getty Images/Michael Steele; p.24 (inset) Getty Images/Andy Lyons; pp. 26-27 Corbis/George Tiedemann; p.29 AAP Image/Anja Niedringhaus.

Table of Contents

Olympic History

The first ever Olympic Games were held in Olympia, in **ancient** Greece. They took place in the year 776 B.C. These ancient games were held every 4 years. They stopped in A.D. 393.

An ancient Greek vase shows men running in an Olympic race.

LET'S EXPLORE MATH

Athens, the **site** of the ancient Olympic Games, has hosted the modern Olympic Games twice. The first time was in 1896 and the second in 2004.

a. How many years passed between these Games?

b. Describe how you solved the problem.

The first **modern** Olympic Summer Games were held in 1896. They took place in Athens, Greece. Only 241 men took part. The 2004 Olympic Summer Games were held in Athens too. This time over 10,000 men and women took part.

The opening ceremony of each Games is always an amazing event.

Many Games

The Olympic Summer Games are not the only Olympic Games. There are also the Olympic Winter Games and the Paralympic Games. The Olympic Winter Games are held every 4 years but in a different year from the Summer Games. The first Winter Games were held in 1924. The Paralympic Games are held in the same year as the Summer Games and the Winter Games.

Host Cities

The Olympic Summer Games are held every 4 years. The Games take place in a host city.

In 1896, Athens, Greece, was the host city for the first modern Olympic Summer Games. Since then, there have been over 20 different host cities.

The 100-meter race at the 1896 Games

Olympic Host Cities

Host cities are announced many years before they hold the Olympic Games. In 2012, the host city is London. In 2016, the host city is Rio de Janeiro.

Tower Bridge in London, England

The Olympic Flame

The Olympic flame goes to the city hosting the Games. The flame is first lit in Olympia, the site of the ancient Olympic Games. Then it travels to the host city by **relay**.

The Olympic flame is passed between relay runners in a special torch.

LET'S EXPLORE MATH

The city of Los Angeles, California, has hosted the Games twice. The first time was in 1932. The second time was 52 years later. What year was this?

The first Olympic torch relay was for the 1936 Olympic Summer Games. The torch traveled nearly 2,000 miles (3,218 km) from Olympia to the city of Berlin in Germany. It visited 7 countries.

The Olympic torch is carried into the stadium in Berlin.

The ancient city of Olympia, Greece

Getting the Torch Around

Different types of **transportation** are used to carry the Olympic torch to the host city. The 1992 Olympic torch traveled 2,675 miles (4,305 km) from Olympia to Barcelona. Of this distance, 926 miles (1,490 km) were covered by bike.

Sporting Venues

A host city must get ready for the Olympic Games. Often, new sports **stadiums** (STAY-dee-uhms) are built. When the Games finish, the stadiums are used for other events in the city.

The National Stadium in Beijing is known as the Bird's Nest. Can you see why?

Under Construction

Building of the National Stadium began in 2003 and finished in 2008.

Stadium Australia

This huge stadium was built for the 2000 Olympic Summer Games in Sydney, Australia. It could seat 110,000 **spectators** (SPEK-tay-ters). At the time, it was the largest Olympic stadium ever built.

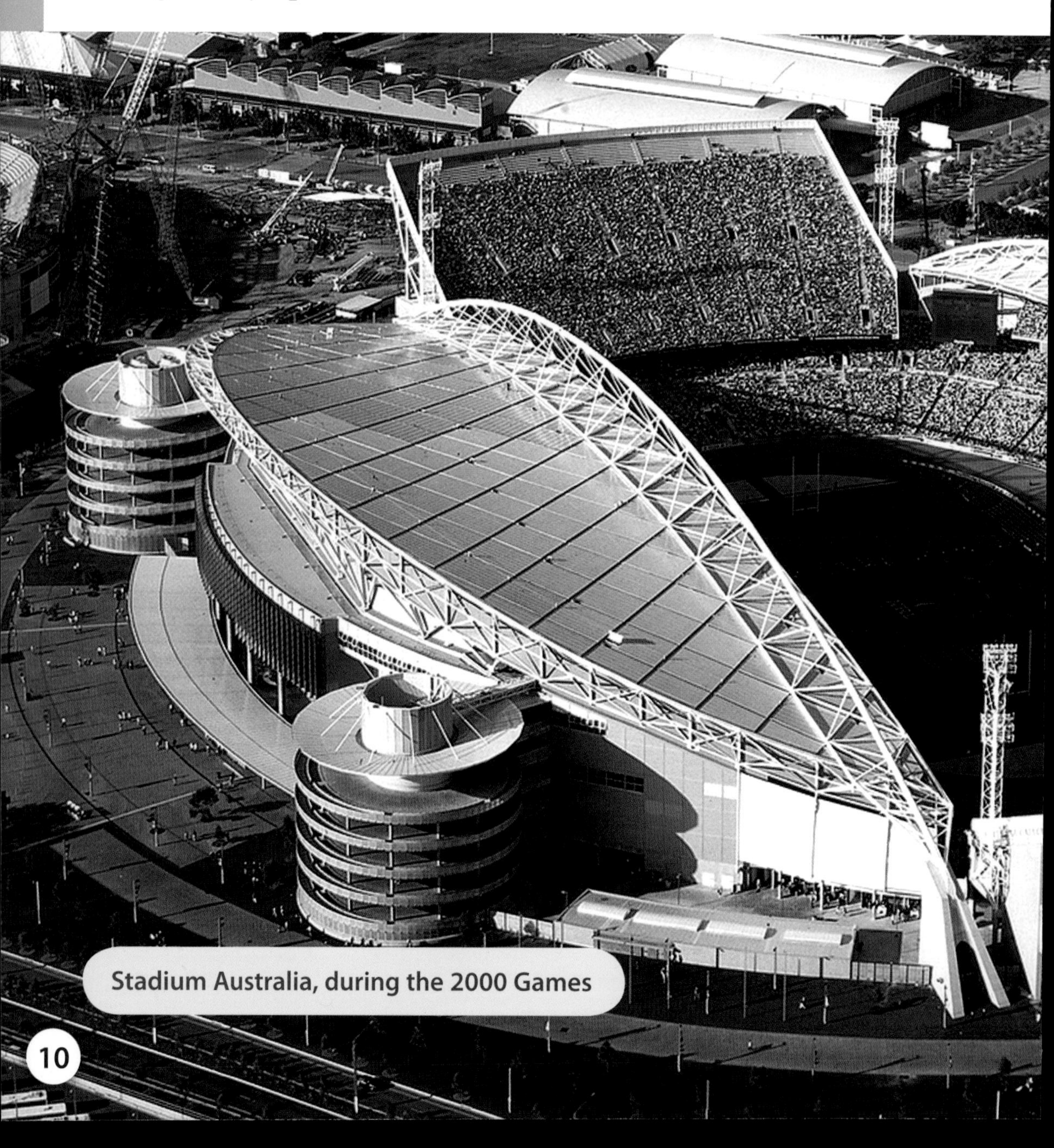

Stadium Australia, during the 2000 Games

It took a long time to build Stadium Australia. The site covers almost 186,000 square yards (155,520 square meters). Its highest point is 190 feet (58 m). That is about the height of a 14-story building.

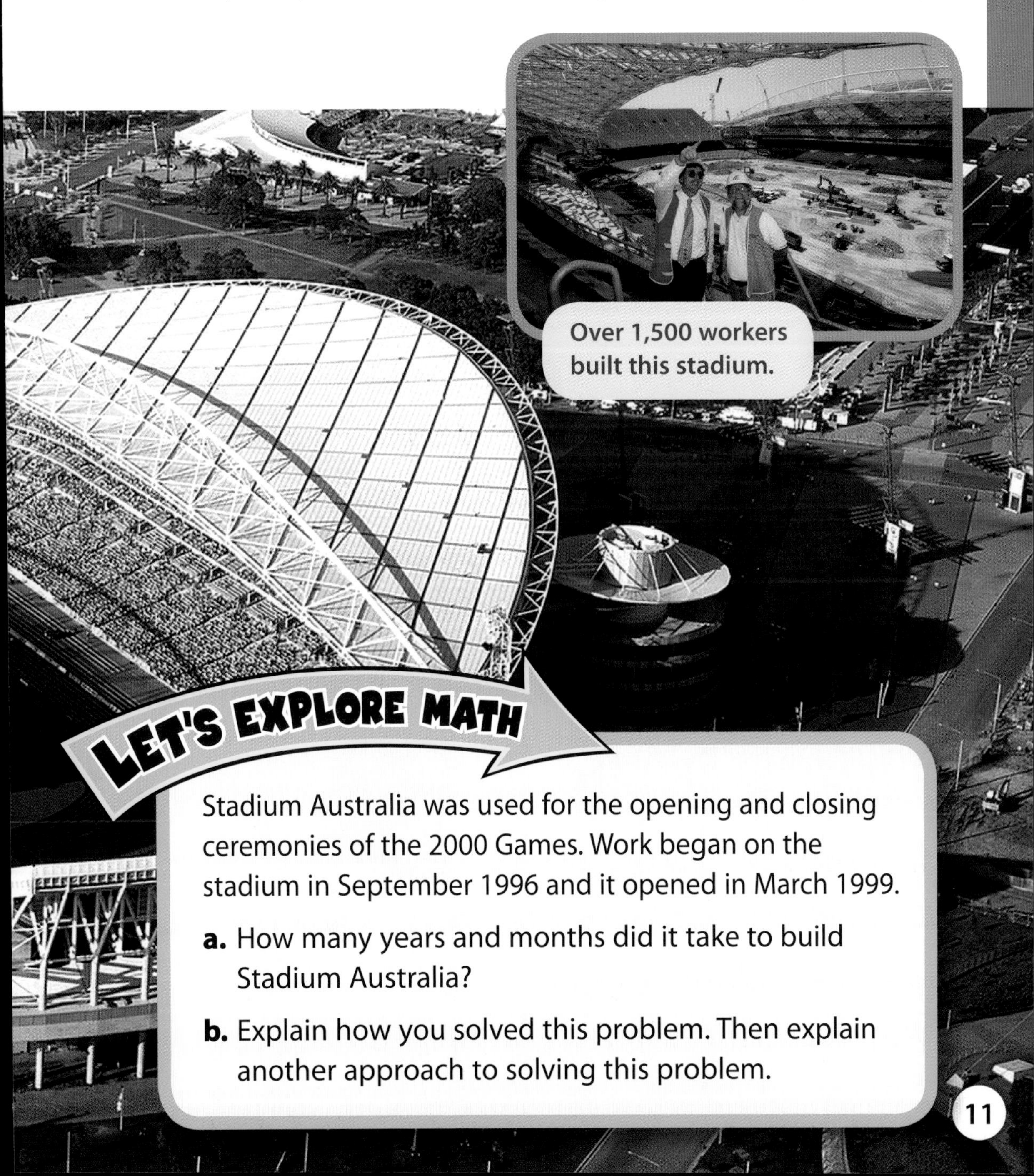

Over 1,500 workers built this stadium.

LET'S EXPLORE MATH

Stadium Australia was used for the opening and closing ceremonies of the 2000 Games. Work began on the stadium in September 1996 and it opened in March 1999.

a. How many years and months did it take to build Stadium Australia?

b. Explain how you solved this problem. Then explain another approach to solving this problem.

Sporting Events

There were 43 events in the 1896 Olympic Summer Games. In 2008, there were 302 events. That is an increase of over 700 percent in the number of events.

In 1996, beach volleyball became an Olympic sport.

Different Sports, Different Events

The 2008 Paralympic Games held events in over 20 different sports. At the 2010 Olympic Winter Games, there were 86 events held in 15 different sports.

Watching the Events

Many people watch Olympic events. The **venues** (VEN-youz) must be big enough to hold a lot of spectators.

At the 1896 Games, the swimming events were held in the Aegean Sea. Nearly 20,000 people watched them from shore. At the 2008 Games, the swimming events were held in an aquatics center. The center seated 17,000 spectators.

Fans at the 2004 Games

Tourism

Many people get to know about the host city by watching the Games. They go to the city or they watch the events on television. In 1992, the Games were held in the city of Barcelona, in Spain. In 1990, before the Games, 1.7 million **tourists** went to Barcelona. In 2005, after the Games, there were 5.5 million tourists. Hosting the Games increases tourism for the host city.

Barcelona

LET'S EXPLORE MATH

The opening ceremony for the 1992 Olympic Games in Barcelona was held on July 25, 1992. A family from the United States went to see it. They left home on July 20 and returned on August 11. How many days were they away?

July 1992

Sunday	Monday	Tuesday	Wednesday	Thursday	Friday	Saturday
			1	2	3	4
5	6	7	8	9	10	11
12	13	14	15	16	17	18
19	20	21	22	23	24	25
26	27	28	29	30	31	

August 1992

Sunday	Monday	Tuesday	Wednesday	Thursday	Friday	Saturday
						1
2	3	4	5	6	7	8
9	10	11	12	13	14	15
16	17	18	19	20	21	22
23	24	25	26	27	28	29
30	31					

Hundreds of thousands of tourists went to Athens for the 2004 Games. More than 4 billion people watched the events on television. In 2005, Greece had more tourists visit the country than before the Games.

Olympics on TV

The 1960 Olympic Summer Games were the first to be shown on television. The Games were held in the city of Rome, in Italy. Millions of people were able to see this city from their own homes.

Food stalls feed the crowds at thc Games. Food was even sold at the ancient Games.

At the 2000 Games, 3,000 bottles of water were sold every day. So were 4,000 ice cream cones and 30,000 hot dogs!

That's Long!

In 1996, a hot dog was made in honor of the Olympic Summer Games in Atlanta, Georgia. It was 1,996 feet (608.3 m) long!

Roads and Transportation

A host city will often improve its transportation systems before the Games start. More people will be in the city, so these systems must be able to **cope**.

In 2004, workers in Athens improved around 450 streets. They fixed over 2 million square feet (185,806 sq m) of pavement. Over 350,000 plants and flowers and 11,000 trees were planted.

A worker fixing pavement and streets before the 2004 Games

About 6 million people traveled by trains and buses to the 2000 Games. That was 15.7 million more rail trips than normal!

Spectators wait in line at a train station during the 2000 Games.

LET'S EXPLORE MATH

On September 23, 1996, it was announced that Sydney, Australia, would host the 2000 Games. The Games opened on September 15, 2000.

a. How many years, months, and days did Sydney have to prepare for the Games?

b. What did you need to know before you could solve this problem?

Atlanta, Georgia, hosted the 1996 Olympic Summer Games. Around 100,000 workers used transportation during the Games. So did more than 2 million spectators.

Huge crowds in Atlanta

Looking After the Athletes

Each host city has an Olympic village. It is where the athletes, coaches, and officials stay. In 2008, the Beijing Olympic village had 42 multistory buildings. Over 16,800 athletes, coaches, and officials stayed here.

The Olympic village in Beijing

The 1976 Olympic village in Montreal, Canada, was built just 0.5 miles (0.8 km) away from the main Olympic venues. The athletes stayed in 4 large, pyramid-shaped buildings. These buildings were 19 stories high.

The Olympic village in Montreal

LET'S EXPLORE MATH

It took athletes only 14 minutes to walk from the Montreal Olympic village to the main venues. If they needed to be at a venue at 12:35 P.M., what time would they have to leave the village? Use the clock to help you.

12:35 P.M.

Food for the Athletes

It takes a lot of food to feed everyone at the Olympic Games. In 2004, the Athens Olympic village served 50,000 meals a day! Plenty of food was needed during the Olympics:

Food	Amount
milk	3,965 gallons (15,000 liters)
eggs	2,500 dozen
fruit and vegetables	330 tons (300 metric tons)
meat	130 tons (120 metric tons)
fish	95 tons (85 metric tons)
bread	25,000 loaves
tomato ketchup	198 gallons (750 L)
drinking water	528,500 gallons (2 million L)

Officials and coaching staff line up for meals in an Olympic village.

LET'S EXPLORE MATH

Look at the table showing the meal times for the dining room in the Olympic village. Use the clock to help you answer the questions.

Breakfast	Lunch	Dinner
5:15 – 9:30 A.M.	12:00 – 2:10 P.M.	5:40 – 9:35 P.M.

a. For how long is breakfast served?

b. For how long is lunch served?

c. For how long is dinner served?

More and More Athletes

Only 311 Greek men took part in the ancient Olympic Games. Today, both male and female athletes take part. They are from 204 different countries.

Athletes from all around the world compete in the Olympic marathons.

In the 1896 Olympics, 241 athletes took part. In 1920, there were 2,626 athletes. There were 9,956 athletes competing in 1992. In 2004, there were 10,625 athletes. And in 2008, over 10,900 athletes competed.

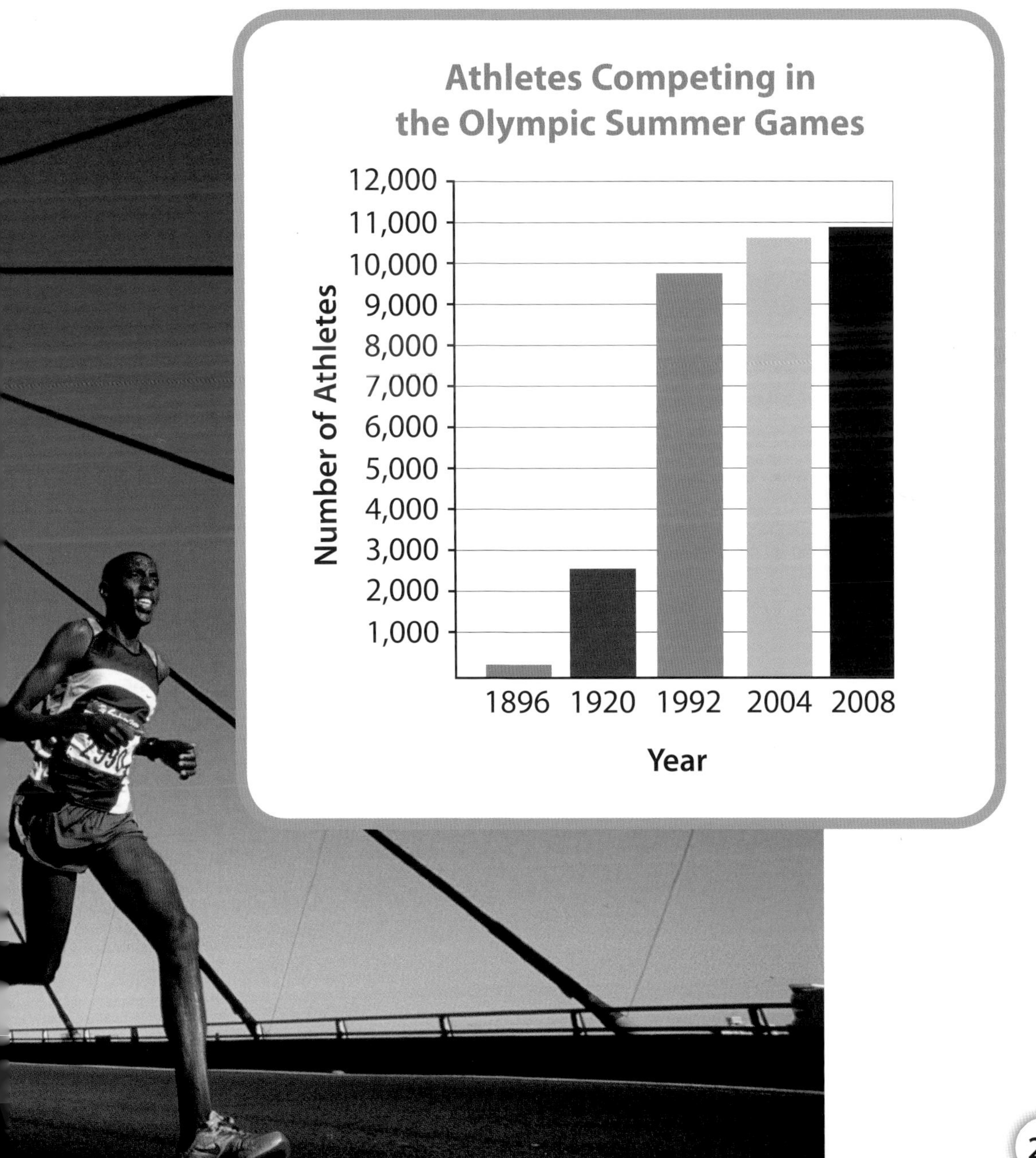

Counting the Cost

Hosting the Olympic Summer Games can cost a city around $10 billion. Just think how much a new stadium must cost to build. But for most host cities, this cost is worth it. The Games bring more tourists to the host city, then and in the future. And the fun of the Games cannot be missed!

Over $500 million was spent building stadiums and venues for the 1996 Atlanta Olympic Games.

Host City Timeline

Olympic Summer Games

Athens, Greece	1896
Paris, France	1900
St. Louis, Missouri	1904
London, England	1908
Stockholm, Sweden	1912
Antwerp, Belgium	1920
Paris, France	1924
Amsterdam, Netherlands	1928
Los Angeles, California	1932
Berlin, Germany	1936
London, England	1948
Helsinki, Finland	1952
Melbourne, Australia	1956
Rome, Italy	1960
Tokyo, Japan	1964
Mexico City, Mexico	1968
Munich, Germany	1972
Montreal, Canada	1976
Moscow, U.S.S.R	1980
Los Angeles, California	1984
Seoul, Korea	1988
Barcelona, Spain	1992
Atlanta, Georgia	1996
Sydney, Australia	2000
Athens, Greece	2004
Beijing, China	2008
London, England	2012
Rio de Janeiro, Brazil	2016

Seeing the Swimming

The Olympic Games are a very popular event for spectators. Jacinta has tickets to one of the swimming sessions, but she needs to find out what time she must leave home so she will not miss any of the races. The session will start at 10:00 A.M.

10:00 A.M.

Jacinta will take public transportation. She will walk to the bus stop, take a bus to the train station, take a train to the Olympic station, and walk from the station to the swimming venue.

- The walk from her home to the bus stop takes 9 minutes.
- The bus ride takes 35 minutes.
- She arrives at the station 6 minutes early.
- Jacinta catches the 8:55 A.M. train.
- The train trip takes 45 minutes.
- It takes Jacinta 20 minutes to walk from the Olympic station to the venue and find her seat.

Solve It!

What time does Jacinta have to leave her home? Use these steps to help you work out your answer.

Step 1: Write the time the session starts, so Jacinta knows what time she needs to be seated.

Step 2: Subtract the number of minutes it will take her to walk from the Olympic station to the venue.

Step 3: Subtract the time the train trip takes, then the waiting time at the station, then the bus trip time, and finally the time it takes her to walk to the bus stop.

Glossary

ancient—very old, belonging to a previous age

cope—manage

modern—present-day, recent

relay—a journey or race where each runner completes part of the distance before handing over an item to the next runner

site—place or location

spectators—people who buy tickets to watch an event

stadiums—sporting arenas

tourists—travelers from another city or country

transportation—moving from one place to another

venues—places, such as stadiums, in which events take place

Index

Answer Key

Let's Explore Math

Page 4:

a. 108 years

b. Answers will vary but could include:

1896 to 1900 = 4 years
1900 to 2000 = 100 years
2000 to 2004 = 4 years
Total: 108 years

Page 7:

1984

Page 11:

a. 2 years and 6 months

b. Answers will vary but could include:

September 1996 to September 1998 = 2 years
September 1998 to March 1999 = 6 months
Total: 2 years and 6 months

Page 14:

22 days

Page 18:

a. September 23, 1996 to September 23, 1999 = 3 years
September 23, 1999 to August 23, 2000 = 11 months
August 23, 2000 to September 15, 2000 = 22 days
Total: 3 years, 11 months, and 22 days

b. Answers will vary.

Page 21:

The athletes would have to leave the village at 12:21 P.M.
12:35 P.M. – 14 minutes = 12:21 P.M.

Page 23:

a. 4 hours, 15 minutes

b. 2 hours, 10 minutes

c. 3 hours, 55 minutes

Pages 28–29:

Problem-Solving Activity

Step 1: The session starts at 10:00 A.M.

Step 2: 10:00 – 20 minutes (walk from station to venue) = 9:40 A.M.

Step 3: 9:40 – 45 minutes (train trip) = 8:55 A.M.

8:55 – 6 minutes (wait at train station) = 8:49 A.M.

8:49 – 35 minutes (bus trip) = 8:14 A.M.

8:14 – 9 minutes (walk from home to bus) = 8:05 A.M.

Jacinta has to leave her home at 8:05 A.M.